# World of Wonder

# MACHINES AND INVENTIONS

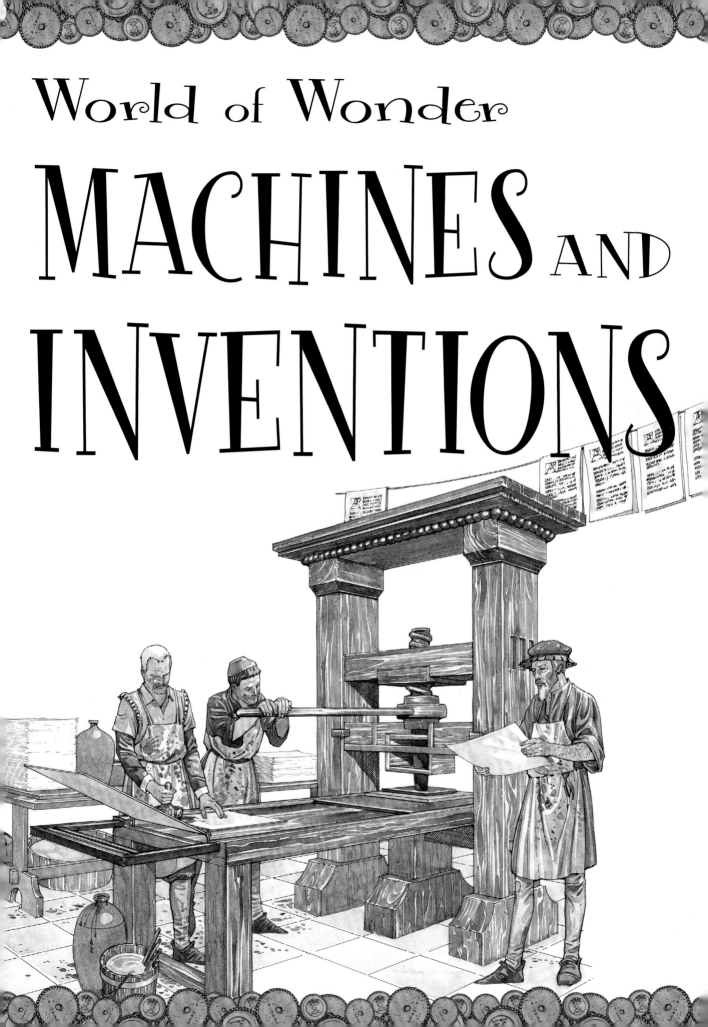

# SALARIYA

Published in Great Britain in 2008
by Book House, an imprint of
**The Salariya Book
Company Ltd**
25 Marlborough Place,
Brighton BN1 1UB

PAPER FROM
SUSTAINABLE
**FORESTS**

*Author:* Ian Graham studied applied physics at the
City University, London. He then took a
postgraduate degree in journalism, specialising in
science and technology. Now a freelance author and
journalist, he has written more than a hundred
children's non-fiction books.

*Artists:* David Antram, Mark Bergin,
Bill Donohoe, Riz Hajdul, John James, Mark
Peppé, Lee Peters, Tony Townsend, Hans Wiborg-
Jenssen, Gerald Wood

*Editor:* Stephen Haynes
*Editorial assistants*: Rob Walker, Tanya Kant

HB ISBN: 978-1-906370-42-8
PB ISBN: 978-1-906370-43-5

A CIP catalogue record for this book is
available from the British Library.

Printed and bound in China.

Visit our website at **www.salariya.com**
for **free** electronic versions of:
**You Wouldn't Want to be an Egyptian Mummy!**
**You Wouldn't Want to be a Roman Gladiator!**
**Avoid Joining Shackleton's Polar Expedition!**
**Avoid Sailing on a 19th-Century Whaling Ship!**

## The world's fastest rocket-plane

NASA's X-15 was the world's fastest
piloted rocket-plane. It made 199
flights between 1959 and 1968,
reaching a top speed of 7,274 kph.

## The world's biggest airliner

The European Airbus A380 is the
world's biggest airliner. It measures 80
metres from wing-tip to wing-tip and
carries 525 passengers on two decks.

# World of Wonder

# Machines and Inventions

by
Ian Graham

Airbus A380

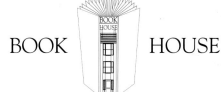

BOOK HOUSE

# Contents

# What are machines and inventions?

Machines are devices that do work. A machine can be as simple as a **lever** or as complicated as a **robot**. An invention is a new way of doing something. Not all inventions are machines.

# What were the very first inventions?

Some of the first inventions were simple tools such as scrapers, knives and axes. They were made from things that people found around them: stones, bones, antlers, branches and other materials. A round stone made a good hammer. A broken stone with a sharp edge was good for cutting.

## How did man make fire?

1.

2.

3.

Thousands of years ago, people learned how to start fires by rubbing pieces of wood together. They could also make sparks by striking together stones called flint and pyrite.

The fire drill (1), flint and pyrite (2), and the fire plough (3) were ancient ways to make fire.

## Barbed fish-hook

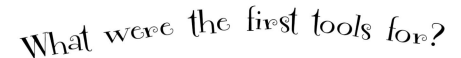

## What were the first tools for?

People used the first tools to provide food and shelter. They used hooks made of bone to catch fish. Stone knives were used to cut up the animals they caught. They cut down trees with stone axes to make shelters.

## Stone-age knives

Flint was used to make razor-sharp cutting tools. Wrapping them with animal skin or fitting a handle made them easier to hold.

## True or False?

Shells were once used as lamps.

Answers on page 31

# When was the wheel first around?

The wheel is one of the most important inventions ever made. It was invented about 5,500 years ago in Mesopotamia. Mesopotamia is the ancient name for the land that is now called Iraq.

## True or False?

This model was found in a tomb in Iraq and dates from 2000 BC.

Answers on page 31

## Types of wheel

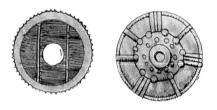

The first wheels were made of wooden planks joined together and then carved into a round shape. These solid wooden wheels were very heavy. Later wheels were made with spokes to save weight.

Ox-cart from ancient Mesopotamia

Wheels made it easy to move heavy loads in carts pulled by horses or oxen. Later, small, lightweight **chariots** carried warriors into battle. Ancient carvings and mosaics show us what these vehicles were like. In the Inca and Maya civilisations, wheels were used only to make push-along toys. They were never used to make real vehicles.

Mesopotamian war chariot

# Who invented measurement?

The ancient Egyptians were among the first people to use systems of measurement. Their units of length were based on parts of the body. The distance from the elbow to the fingertips was one **cubit**.

## Measuring the pyramids

Egyptian pyramid builders had to make sure the ground was level before they could start building. Here is one way they may have done it.

First they measured out a huge grid of squares on the ground. Then they cut grooves between the squares and filled the grooves with water. Because all the grooves were joined together, the water settled to the same height in all the grooves. They marked the water level inside each groove, then they dug all of the ground down to this level, one square at a time.

Grooves

Grid of squares

# Measuring time

The **sundial** is one of the oldest inventions for telling the time.

When the sun shines, the pointer (called a gnomon) casts a shadow on the dial. As the earth turns, the sun appears to move across the sky. This makes the shadow of the gnomon move as well. The shadow points to the numbers on the dial, which show the time of day.

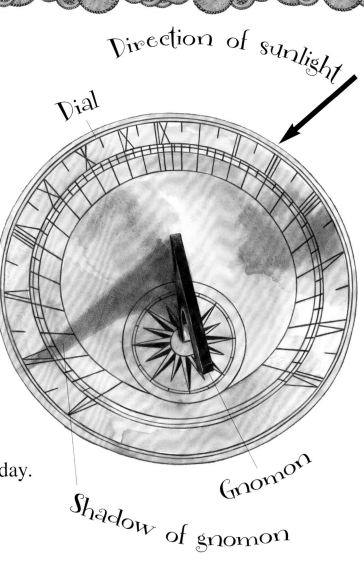

Direction of sunlight

Dial

Gnomon

Shadow of gnomon

Pendulum clock

Pocket watch

The first clocks powered by weights or springs were built in Europe in the 1300s. Watches came along 200 years later. At first they were carried in a pocket. Wristwatches became popular in the 1920s.

# What were the first forms of defence?

The first forms of defence were the same spears and clubs that people used to hunt animals for food. Later, people made weapons specially for fighting, and formed armies for defence.

In the Middle Ages, castles were built across Europe. At first they were made of wood, but later stone was used. Stone castles are so strong that many are still standing today.

But even the strongest castle could still be captured by siege.

Crane

Treadmill

Outer wall

More than 2,700 years ago, the Assyrian army was the first to be fully equipped with weapons made of iron.

Shield

Archer

Helmet

# Building a medieval castle

Well

Corner tower

Dungeon

In a siege, enemy soldiers surrounded the castle so that no-one could get in or out. The defenders had to surrender when they ran out of food. The enemy might also try to break down the castle walls using machines that threw huge rocks.

This toilet was built in about 2500 BC in what is now Pakistan.

# How did people keep clean?

**B**athing became common in the days of ancient Rome, about 2,000 years ago. Instead of using soap, Romans scraped their skin clean with a tool called a **strigil**.

Ancient Rome had many public baths. The Baths of Caracalla, shown here, were the second biggest.

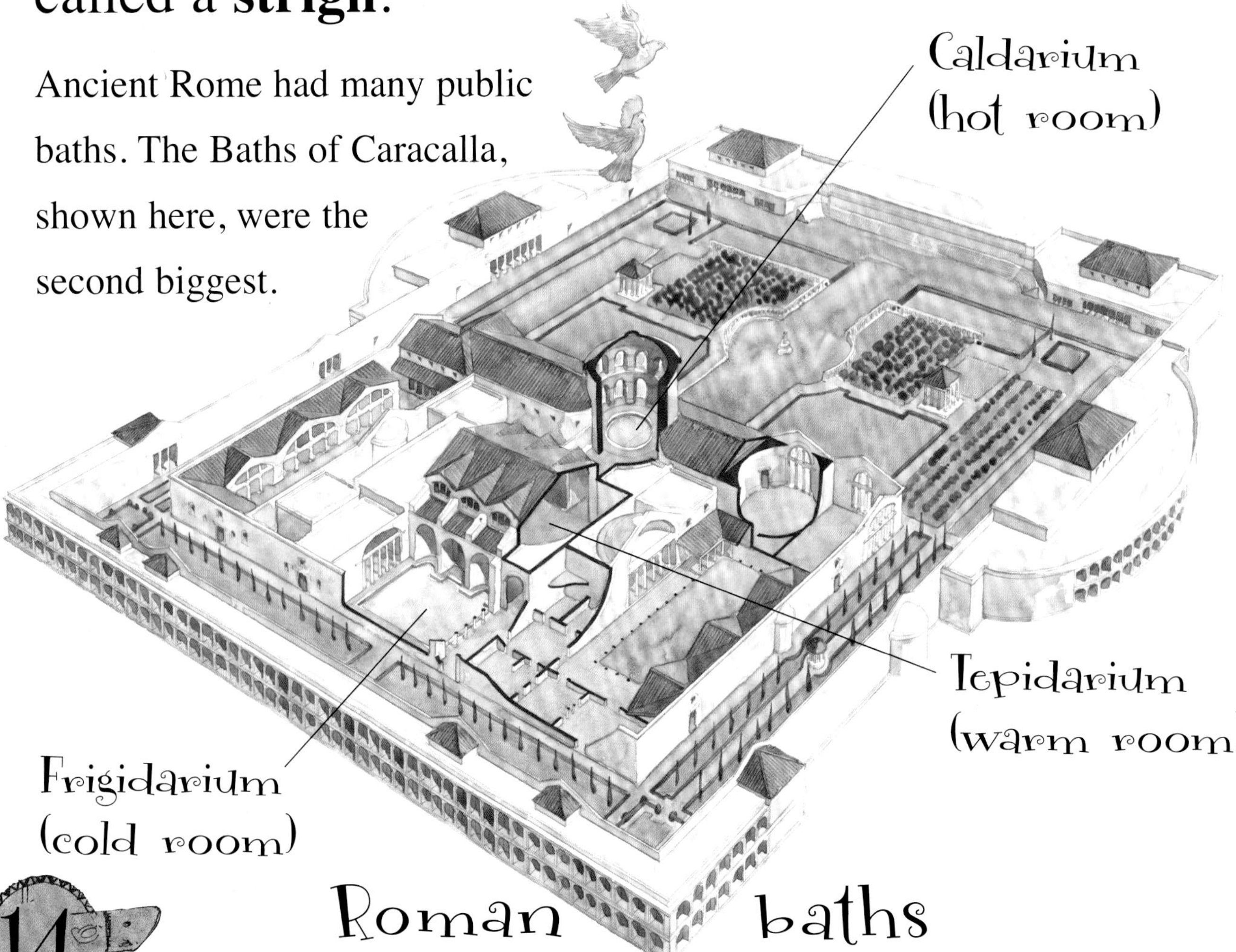

Caldarium (hot room)

Tepidarium (warm room)

Frigidarium (cold room)

Roman baths

## Minoan bath

The oldest known bathtubs are about 4,000 years old. They were found on the island of Crete in the Mediterranean, where the Minoan people lived. Their baths had no taps – they were filled from jugs. Minoan toilets were flushed with a jug of water, too.

By the 1890s, grand houses had their own flush toilets, known as 'water closets'. Pulling a handle sent water rushing through the toilet to clean it. We still use the same method today.

## Victorian water closet
### c.1890

The oldest Chinese writing used picture symbols.

# Who first used writing?

The Sumerian people in Mesopotamia (ancient Iraq) were the first to use writing. They used it to keep records of the farm produce they bought and sold. They wrote by drawing pictures in soft clay with the end of a reed. This picture-writing developed into *cuneiform*, a type of writing that uses wedge-shaped symbols.

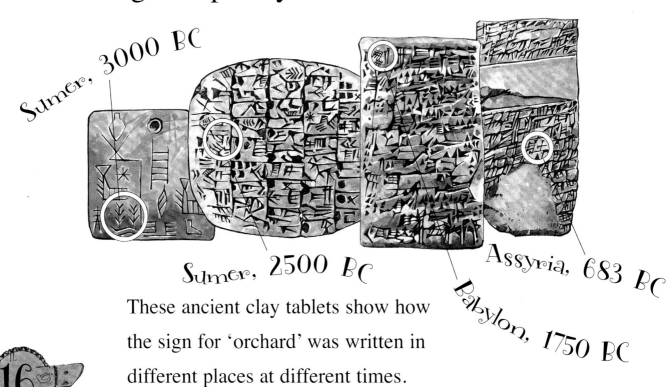

Sumer, 3000 BC

Sumer, 2500 BC

Assyria, 683 BC

Babylon, 1750 BC

These ancient clay tablets show how the sign for 'orchard' was written in different places at different times.

## When was printing invented?

The oldest known printed book is a Chinese book called the *Diamond Sutra*. It is dated 11 May AD 868. It was printed from wooden blocks on which all the symbols and pictures were carved by hand. Later, the Chinese invented movable type.

Movable type uses separate letters made of metal that can be put together in different ways to make words. The same metal letters can be used again and again.

In about 1450, Johannes Gutenberg built the first printing press using movable type.

Lever

Screw

Ink

Press

Gutenberg's printing press

Roman watermills

At least 2,000 years ago the Greeks and Romans used the power of flowing water to turn heavy millstones to grind grain into flour.

Grain goes in

Water trough

Millstones

Flour comes out

Gears

Waterwheel

# Who first harnessed the power of nature?

The first machines that used the power of nature were built by the ancient Greeks and Romans. They started using waterwheels in about 100 BC. Flowing water turns a waterwheel by pushing against paddles or scoops around its edge.

# Newcomen's steam engine

English engineer Thomas Newcomen built the first working steam engine in about 1712. Newcomen's engines were used to stop mines from flooding by pumping water out of them.

Beam

Pump rod

Mine shaft

Boiler

Cylinder

## Who made the first light bulb?

The first successful light bulbs were made in the 1870s by the British inventor Sir Joseph Swan and the American Thomas Edison.

## True or False?

Michael Faraday invented a **generator** in 1831.

? ? ?

Answers on page 31

# When did we start to travel fast?

Steam trains were invented in the early 1800s. Before that, railway wagons were pulled by horses. Steam engines were a great success and railways quickly spread across the world. Cars and planes came later, after smaller, lighter petrol engines had been invented.

The first powered flight took place on 17 December 1903, at Kitty Hawk in North Carolina, USA. Orville Wright flew the plane, which he and his brother Wilbur had built.

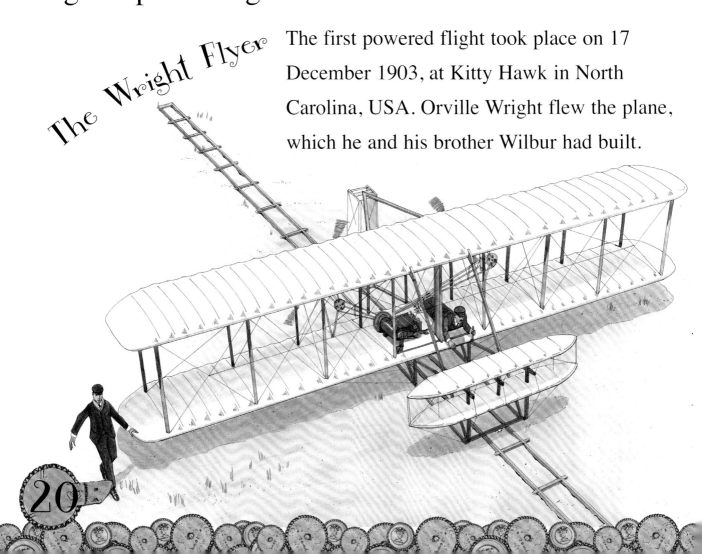

The Wright Flyer

The first petrol-engined car was built in 1885. Early cars were very expensive. Then in 1908 Henry Ford made the first car that ordinary people could afford to buy: the Model T. More than 15 million of these were made between 1908 and 1927.

Ford Model T

## Why is the Rocket famous?

### True or False?

The first steam-powered vehicle ran on rails.

? ? ? ?

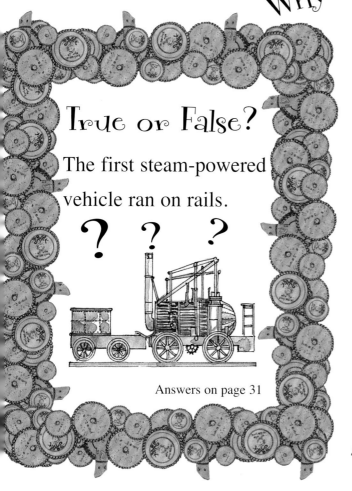

Answers on page 31

The *Rocket* is famous because it won a competition held in 1829 to find the best railway locomotive.

# Thrust SSC

Body shell

### True or False?

The Lockheed Blackbird is the fastest aeroplane.

? ?

Answers on page 31

*Thrust SSC*'s body shell is made from aluminium, carbon-fibre and titanium. This makes it both light and strong.

# What's the fastest thing on wheels?

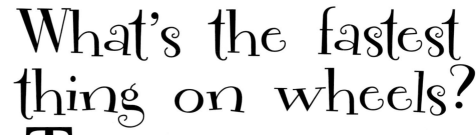

**Rolls-Royce jet engine**

The fastest car so far is called *Thrust SSC*. In 1997 this jet-powered car set a world land speed record of 1,228 kph (763 mph).

Air intake

*SSC* stands for '**supersonic** car'. *Thrust SSC* was the first car able to go faster than the speed of sound.

# How did modern medicine start?

For thousands of years, doctors tried to cure their patients with herbs and spells. Over the past few hundred years, scientists slowly learned how the body works, and what really causes diseases.

## Making medicine safer

**Anaesthetics** were discovered in the 1840s. Doctors and dentists found that patients didn't feel pain when they breathed a gas called ether.

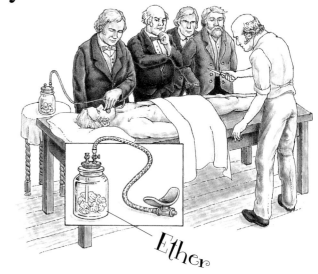

Ether

After 1865, Joseph Lister sprayed operating theatres with **carbolic acid** to kill germs.

Carbolic spray

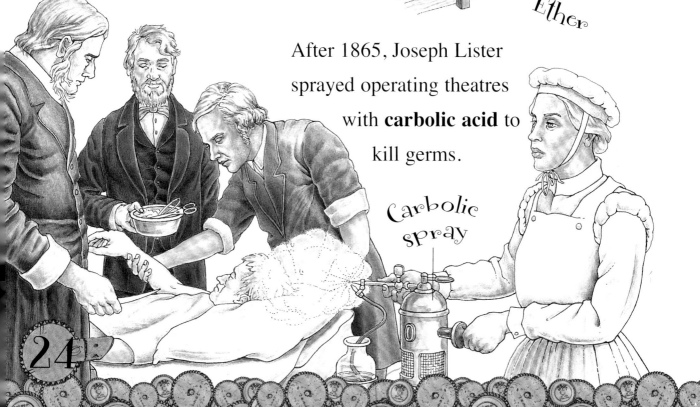

## Louis Pasteur

Edward Jenner discovered **vaccination** for smallpox in 1796. In the 1880s, Louis Pasteur found a vaccine for another killer disease, rabies.

## True or False?

You can grow skin in a laboratory.

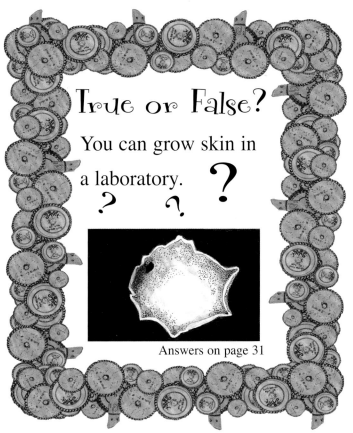

Answers on page 31

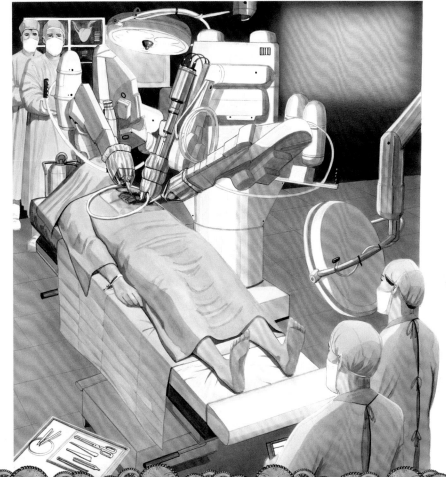

## Robot surgery

Today, some surgical operations are carried out by medical robots controlled by surgeons. The surgeons watch the operation through a video camera on one of the robot's arms.

# Who invented modern communications?

Telephones, radio, television and the Internet help us communicate with other people and learn about world events as soon as they happen.

*Mouthpiece*

*Early telephone*

The first successful telephone was invented by Scotsman Alexander Graham Bell in 1876, when he was still in his twenties.

*Marconi's radio transmitter, 1895*

The Italian inventor Guglielmo Marconi did not discover radio waves or build the first radio equipment, but he did more than any other inventor to make radio communication popular and successful.

# John Logie Baird

Scotsman John Logie Baird made the first television set in 1926. He made his television from scrap materials including a hat box and a biscuit tin!

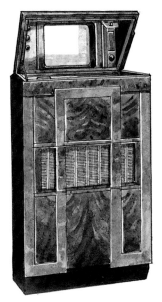

TV set, 1930s

## How did the Internet start?

In the 1960s, scientists and engineers in the USA linked computers in different places together to form a **computer network**. More networks were set up in other countries. In 1983, computer networks all over the world were linked together to form the Internet.

# What will they think of next?

**S**ome of the machines and inventions we use today are amazing. What sort of wonders might inventors think up in future? Tomorrow's life-changing inventions may come from advances in computers, robots and genetics.

## What is genetic engineering?

All living cells have a substance called **DNA** inside them. DNA contains instructions that tell the cells how to grow. It's divided into tiny pieces called **genes**. **Genetic** engineering is a way of changing a plant or animal's DNA.

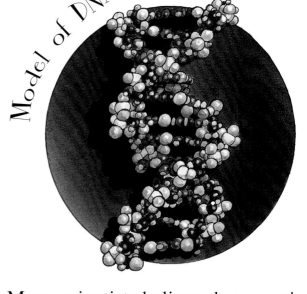

Model of DNA

Many scientists believe that genetic engineering can create better farm crops. Improved crops could produce more food, and have fewer diseases than ordinary crops. But some think it's wrong to interfere with nature.

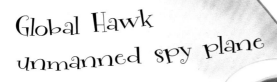
Global Hawk
unmanned spy plane

# Robots of the future

There are already robot planes that fly
without a pilot inside, industrial
robots in factories and walking robots
that look like people. Future robots
may help around the home.

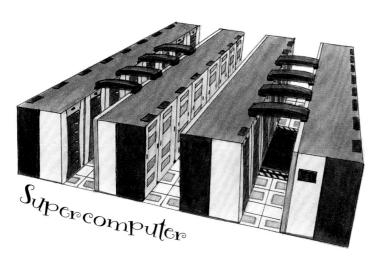
Supercomputer

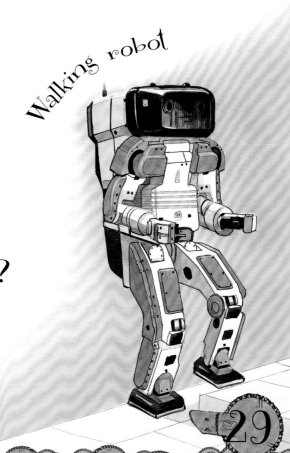
Walking robot

# What are supercomputers?

Supercomputers are the world's fastest
computers. They tackle the most difficult
problems, such as studying the world's
weather and **climate**.

# Useful words

**Anaesthetic**   A drug that stops people from feeling pain.

**Carbolic acid**   An old-fashioned name for phenol, a poisonous chemical that was once used to kill germs in hospitals. It is still used for making medicines.

**Chariot**   A fast, light horse-drawn cart used in wartime and for racing.

**Climate**   The average weather conditions over a long time.

**Computer network**   A number of computers linked together.

**Cubit**   An ancient unit of measurement: the distance from the elbow to the outstretched fingertips.

**DNA**   A substance found in all living cells that contains instructions telling the cells how to grow. DNA stands for 'deoxyribonucleic acid'.

**Generator**   A machine for producing electricity. A dynamo on a bicycle is a kind of generator.

**Genes**   Pieces of DNA that pass on characteristics from one generation of living things to the next. You look the way you do because of genes you inherited from your parents.

**Genetic**   Having to do with genes.

**Lever**   A bar that rests on a prop (called a fulcrum). By moving one end of the bar you can move a heavy object resting against the other end.

**Robot**   A machine that can be programmed to do a particular kind of work automatically.

**Strigil**   A tool with a curved bronze blade for scraping oil and dirt off the skin.

**Sundial**   A device for telling the time. A pointer (called a gnomon) casts a shadow on the dial, and the shadow moves around the dial as the sun moves across the sky.

**Supersonic**   Faster than the speed of sound.

**Vaccination**   Giving weakened or dead germs to a person or animal to help the body produce its own defences against a disease.

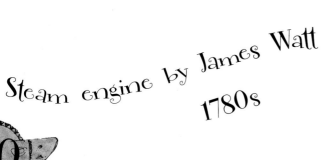

Steam engine by James Watt 1780s

# Answers

**Page 7** **TRUE!** At least 20,000 years ago, people made lamps from natural objects such as shells. The shell was filled with oil or fat. The simplest form of wick was made from rush.

**Page 8** **FALSE!** The model cart did come from a tomb in Iraq, but it dates from 3000 BC. That makes it 5,000 years old.

**Page 17** **TRUE!** Johannes Gutenberg printed a magnificent Bible in the 1450s. It is sometimes called the Mazarin Bible, because one of the best copies was found in the library of Cardinal Jules Mazarin in Paris in 1760.

**Page 19** **TRUE!** In 1831, the English scientist and inventor Michael Faraday discovered that he could make an electric current flow in a coil of wire by moving a magnet near the coil. Modern generators use magnetism to make electricity in the same way.

**Page 21** **FALSE!** The first steam-powered vehicle was a three-wheeled tractor. It was built in 1769 by a French army engineer called Nicolas Cugnot. It was designed for pulling cannons.

*Benz three-wheeler*

The first petrol-powered vechicle, invented in 1885 by Karl Benz

**Page 22** **TRUE!** The Lockheed SR-71 Blackbird is the fastest aircraft that can take off from the ground. In 1976 it reached a speed of 3,529 kph. That's more than three times the speed of sound. Rocket-planes (see page 2) are even faster, but they can't take off by themselves – they have to be launched from a larger plane.

**Page 25** **TRUE!** Skin has been grown in laboratories since the 1970s. It is used for covering burns, when there isn't enough of the patient's own skin to use. Artificial skin is made using substances taken from sharks!

# Index

(Illustrations are shown in **bold type**.)

A very small robot used to
inspect factory pipelines